Dishonour of the Crown

DISHONOUR OF THE CROWN

THE ONTARIO RESOURCE REGIME IN THE VALLEY OF THE KIJI SÌBÌ

Paula Sherman
Ardoch Algonquin First Nation

Foreword by Leanne Simpson

ARBEITER RING PUBLISHING • WINNIPEG

Arbeiter Ring Publishing
201E-121 Osborne Street
Winnipeg, Manitoba
Canada R3L 1Y4
www.arbeiterring.com

Printed in Canada by Transcontinental Printing
Cover by Michael Carroll

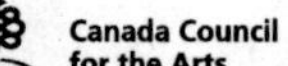

Canadian Heritage
Patrimoine canadien

With assistance of the Manitoba Arts Council/Conseil des Arts du Manitoba.

We acknowledge the support of the Canada Council for our publishing program.

ARP acknowledges the financial support to our publishing activities of the Manitoba Arts Council/Conseil des Arts du Manitoba, Manitoba Culture, Heritage and Tourism, and the Government of Canada through the Book Publishing Industry Development Program (BPIDP).

Printed on 100% recycled paper.

Library and Archives Canada Cataloguing in Publication

Sherman, Paula

Dishonour of the Crown : the Ontario resource regime in the valley of the Kiji Sìbì / Paula Sherman.

Includes bibliographical references.
ISBN 978-1-894037-36-5

1. Ardoch Algonquin First Nation. 2. Algonquin Indians—Ontario—Government relations. 3. Algonquin Indians—Land tenure—Ontario, Eastern—History. 4. Uranium mines and mining—Environmental aspects—Ontario, Eastern. I. Title.

E99.A35S54 2008 323.1197'30713 C2008-906547-6

I would like to express my gratitude to all the people who donated their time to this project. I would particularly like to thank Chris Furgal and James Wilkes from Trent University, and Dr. Linda Harvey for the information and sources they passed onto me during the research phase. In addition I would like to thank Leanne Simpson, Robert Lovelace, Mireille Lapointe, and Gloria Morrison for the time they spent reading the draft.

A special thank you to Leanne Simpson for writing the foreword to the book and to all of those who supported our resistance.

CONTENTS

FOREWORD

Leanne Simpson

OVER ONE HUNDRED YEARS AGO, THE FEDERAL GOVERNMENT expropriated Mississauga lands and water to complete the Trent-Severn Waterway, a system of locks and canals linking Lake Ontario and Georgian Bay. The waterway was heralded by settler governments as a major advance that would open up access to markets in the south, strengthening the economy to the benefit of all Ontarians and, indeed, all Canadians.

But not for the Mississauga. The ancient travel route was our river highway, leading to a network of travel and trading routes to the north and the west—an ancient travel route that now was physically colonized. No longer could the lifeblood of our Mother run unobstructed through the veins of our lands. No longer could that lifeblood nurture and cleanse our land. Our Elders knew the impact of the waterway would be devastating to our lands, lakes and rivers and especially to our manoomiin, our wild rice. In an attempt to mitigate these impacts, we turned to our friends and our neighbouring nation, the Omàmìwinini. One of our Elders gave manoomiin seeds from Rice Lake to Omàmìwinini Elders who planted those seeds in the Mississippi River, near Ardoch Ontario. The Ardoch community has nurtured and protected those seeds through the past century, so that each fall our community members can harvest our shared manoomiin.

My community would not have manoomiin if it were not for the forward thinking of Mississauga Elders and their relationship with Omàmìwinini Elders. And we would not have manoomiin without the careful watch and active protection of the Omàmìwinini people. My ancestors, like the Omàmìwinini, did not bank capital or store surplus goods to ensure our economic security or political stability. We banked relationships. In tough times, we did not rely on accumulated wealth for survival, but upon the generosity, knowledge and the nurturing of our friends and neighbouring nations. Our security and stability as a nation, indeed our survival as a nation, was ensured by practising good relations, and by promoting healthy relationships with other Indigenous nations. And so given this historic relationship, it was with heavy heart that I watched the events of the summer of 2007 unfold.

The Omàmìwinini of the Ardoch Algonquin First Nation are part of the larger Omàmìwinini nation, a nation that has never surrendered any of its territory to colonial governments. For the past 400 years, Omàmìwinini have struggled to maintain their sacred relationships within the valley of Kiji Sìbì, the Ottawa River, as a nation illegally occupied by the Canadian state, in spite of the continual imposition of foreign social and political structures. This is not the relationship the Omàmìwinini envisioned when they entered into a wampum agreement with the English and the French in the 1700s. The Friendship wampum acknowledged and continues to acknowledge the nationhood and sovereignty of the Omàmìwinini. It acknowledges the spiritual nature of the relationship the Omàmìwinini have to their land, and it reminds the Omàmìwinini of the responsibility they have to protect the land, the people and the health and welfare of the plants and animals that encompass the Omàmìwinini's web of life.

The Omàmìwinini have a long history of engaging in relationships that promote Pimaatisiwin, the good life, and that promote environmental sustainability. In pre-colonial times, the Omàmìwinini had an ongoing relationship with the Haudenosaunee Confederacy, which the Omàmìwinini called "Our Dish" or Gdoo-naaganinaa. The agreement was designed to protect the areas of

Omàmìwinini they share with the Haudenosuanee Confederacy by acknowledging each other's separate jurisdiction over a shared territory and emphasising that both nations had and continue to have the responsibility to maintain and protect the ecological integrity of the territory, or Our Dish.

The story of the Omàmìwinini and their responsibility to protect their territory from the exploits of uranium prospecting and ultimately uranium mining reached the television screens of ordinary Canadians in the summer of 2007. The Omàmìwinini of Ardoch First Nation once again joined the legions of Indigenous Peoples who have been forced to mobilize to protect their lands. Once again, they joined the longest running resistance movement in Canadian history, including the mobilizations at Ipperwash, Burnt Church, Kanasatake, Cross Lake, Listuguj, Sun Peaks, Gustafsen Lake, and the Six Nations to name just a very few. They blocked access to a uranium-prospecting site as a last resort to protect their lands and to prevent Frontenac Ventures from illegally prospecting on their territory.

Their stories will live on in the oral tradition of the Omàmìwinini and the Mississauga. We will tell our children of the great lengths the Omàmìwinini went to in order to protect their lands and the health of all Ontarians. I will tell my children how proud I felt of the Omàmìwinini in the boardrooms, in the jails, and at the blockade sites, and in the countless research papers, presentations and workshops they have given us all. However, it is not enough to just feel proud; we must use their example to inspire and ultimately to *act.* We must support our friends and allies in times of trouble. It is how we survive. It is how we will flourish once again.

What follows began as the commitment of one Family Head on Ka-Pishkawandemin (the traditional government of Ardoch First Nation)—a commitment to living up to her responsibility to her family, her community, her nation and her land. Paula Sherman originally wrote this manuscript as an educational paper for those interested in the potential impacts of uranium exploration and mining in her own territory. However, it has grown into something much larger than education about these impacts. It is a call to action.

Dishonour of the Crown is a critical intervention not just for citizens of Indigenous nations but also for Canadians. Our current environmental crisis is a direct result of short-term environmental decision-making that privileges the wealth of a handful of corporations over the long-term health and wellness of the citizens of Indigenous nations and Canada. It is high time that we approach environmental decision-making on Turtle Island with precaution, with the best of both western science and Indigenous

Knowledge and with the higher vision of promoting sustainability rather than continuing to make decisions that promote disconnection, destruction, and greed.

Miigwech
Leanne Simpson (Mississauga)

INTRODUCTION TO THE ISSUE OF RESOURCE EXTRACTIONS ON INDIGENOUS LANDS

OVER THE PAST FOUR HUNDRED YEARS, OMÀMÌWININI (Algonquin)[1] people have struggled to maintain our long-standing relationships within the valley of the Kiji Sìbì as a result of French, English, and Canadian colonialism and the implementation of foreign social and political structures.[2] These structures have led to policies that enabled Europeans to profit from our lands and resources while economically and politically marginalizing us to the fringes of our own territory. These policies not only worked to jeopardize our autonomy and jurisdiction, they also attempted to colonize our minds and transform our perceptions of our relationships with the land and even with each other.[3] The implementation of colonial policies of assimilation led to a loss of spirituality and connections to the land in many instances. What's more, the internalized guilt and self-loathing that accompanied cognitive colonization often limited the ability of Indigenous peoples to oppose development projects within Indigenous homelands.

Through a process of re-education to Anishinaabe philosophy and values over the past thirty years, the people of Ardoch Algonquin First Nation have begun to rebuild our relationships within our homeland and to reclaim our autonomy within our community lands. A large component of this process required the development of programs that would build self-esteem and self-confidence so it would be possible to implement community structures that would promote the restoration of our relationships with the land. This has been a difficult process for the community given the fact that Ardoch is a historic community of Omàmìwinini families who are not organized under the Indian Act. It is extremely difficult for historic communities to develop or implement programming when they do not receive education or community funding from the government. All of the programs and services that have been implemented within the community have been accomplished through fundraising activities.

In spite of these difficulties, Omàmìwinini families have begun the process of envisioning an alternative relationship within our homeland that is based upon the Anishinaabe theory of Pimaadiziwin, which situates human beings within a collective that includes the Natural World and all other parts of Creation. Elders such as William Commanda, Harold Perry, Edna Manitowabi, Shirley Williams, and Doug Williams have made real differences in the lives of various Algonquin people, which have facilitated a deeper connection to the land and waterscapes within the Kiji Sìbì. While some Omàmìwinini people continue to find it difficult to make that connection, others have entirely transformed their lives as a result of the recognition that they have responsibilities as human beings within our homeland. Opposition to development has been one of the outcomes of this re-envisioning of Algonquin relationships with the land.

This process of re-envisioning ourselves within the landscape was first put to the test thirty years ago when the Province of Ontario sold away our manoomiin beds for one dollar to a commercial interest. Ardoch Algonquin people stood up in opposition to this decision and were backed by a large contingent of private landowners, environmentalists, and other Indigenous peoples. The manoomiin was a gift from our Mississauga relatives in the nineteenth century, who came to our assistance in a time of starvation. Over the past one hundred and fifty years, those seeds have expanded to cover four beds of rice which have been protected and cared for by the Perry family for generations. This manoomiin nourishes our bodies and spirits and is very important to our survival as a distinct people.

It was the manoomiin, in fact, and our responsibility to protect it, that brought Ardoch Algonquin people out of the slumber we were in and pushed us toward the adoption of a community resolution on development. This process was crystallized in the *Guiding Principles*[4] that were developed during this period. These principles provided guidelines the community would use to rebuild our relationships within our homeland:

> Algonquin people are the first people of the Kiji Sìbì watershed and therefore have a special responsibility to ensure that the land is cared for. Algonquin people can look to no other place in the world to find their origins. Kijimanito created Omàmìwinini (Algonquin) in this valley to give them life and purpose. The creatures with whom we share this valley are our closest relatives. All that is Algonquin, our culture, spiritual practices, language, governance, honour and relationships is the 'story' of this land.[5]

The Guiding Principles provided the foundation for the direct actions that developed around the issue. For the first time in a long time, Omàmìwinini people were able to come together in support of the land and the waterscapes in our homeland. With the help of our neighbours, we were successful in defending the manoomiin beds from commercial interests.

With this success under our belts, the community went on to launch court cases that have led to the establishment of hunting and harvesting rights for communities that are not registered under the Indian Act. The community more recently successfully articulated its position with the Ministry of Natural Resources with respect to our right to build a cultural community centre on our lands near Ardoch. The community also established a charitable organization in 2003 to raise the funds necessary to provide cultural programs and social services to community members. The hope was that with the construction of the cultural centre, we would have the physical capacity to provide cultural and linguistic programming that would facilitate the development of spiritual connections to the lands and waterscapes among our children and youth.

We had just raised funds and begun the process of offering cultural and youth programming when we received the news from Gloria Morrison, a non-Algonquin neighbour, in November of 2006 that our community lands may have been staked for mineral exploration. This was confirmed by members who came back with evidence in the form of photographs that detailed the staking. Overall, 30,000 acres had been staked, 26,000 of which was on Algonquin land that had not been surrendered to the Crown through any treaty or negotiations process.[6] The Morrisons, who owned one hundred acres, discovered that 65 percent of their private lands had been staked. They had exhausted every avenue trying to resolve the issue but came to the realization that there was no provincial or federal protection available because the Ontario Mining Act permitted exploration companies the right to explore for minerals on private and public lands. In the end, the Morrisons came to us hoping that we could provide assistance because the private lands they occupied were part of the larger Algonquin homeland that was never ceded and was under a contemporary comprehensive land claims process.

When our correspondence with the Province and the exploration company was ignored, we established an alliance with our non-Algonquin neighbours whose lands had also been staked and began to develop a collective strategy to deal with the issue. We agreed that the alliance would focus its efforts on the concerns of both Algonquin people and private landowners so we could cover all aspects of the issue. Out of this small local alliance a much broader alliance developed across the region and people organized themselves as community collectives around the issues related to the uranium cycle. The first of these organizations was the Community Coalition Against the Mining of Uranium (CCAMU). CCAMU was soon joined by the Ottawa Coalition Against the Mining of Uranium (OCAMU), and many others. All of these community organizations focused on the issue as it related to private landowners, while also supporting our position as Algonquin people.

Our position as Omàmìwinini people has been that this project was approved without any notification or consultation as is required under the Canadian legal system. Lawyer and activist Christopher Reid offers that: "Over the years, the Supreme Court of Canada and many lower courts have ruled that government and private companies have a duty to consult with Aboriginal people whenever Crown decisions or actions have the potential to adversely affect Treaty or Aboriginal rights. Government decisions about public lands and waters within which Aboriginal peoples have an interest can also trigger the duty to consult."[7] Reid goes on to say, "This duty to consult arises from Section 35 of The Constitution Act, 1982, which recognizes and affirms the 'existing Aboriginal and Treaty rights of the Aboriginal people of Canada.'"[8] Clearly, there was a duty to consult, yet that process was never acknowledged nor attempted by the Crown in this case.

Aside from the issue of no prior notice, and no consultation whatsoever, we were startled to discover that there would be no environmental assessment of the work Frontenac Ventures (FVC) proposed to carry out on our lands. This was particularly disturbing to us since this was a uranium exploration project, which we felt carried with it significant risks to the land, water, animals, and human beings. When additional statements and correspondence went unanswered, we moved onto the Robertsville mine site to prevent access to our lands. We held the site until October of 2007. That action led the Province to respond by calling for a meeting between us and FVC. At this meeting FVC revealed their exploration plans and offered to pay us $10,000 for the right to unfettered access to the lands in question. We declined their offer, and left. The next day they filled a $77 million lawsuit in the Ontario court system as well as a motion for an injunction to have us arrested and removed from the site.

As a result of this action on the part of Frontenac Ventures, the alliance broadened to include people from all walks of life and many environmental organizations. Other Indigenous peoples also came to the site to show their support and offer their assistance. The Ontario Provincial Police (OPP), who had been sent in to deal with issues of public safety, took a position early in the process that they would not make any decisions about colour of right[9] (the right of ownership), which made it very difficult for the injunction to be enforced once it was issued in August. The OPP continued to facilitate public safety but made no move to arrest anyone or remove people from the site.

When the injunction motion was heard before Justice Gordon Thompson's court, we attended briefly and then withdrew from the legal process, citing the words of Justice Sidney Lynden in the Ipperwash Inquiry who offered in his decision that injunctions and court processes were impediments to resolving such conflicts.[10] We agreed with Justice Lynden's perception that conflicts over disputed lands needed a political solution.[11] That was the reason for moving onto the site in the first place: to bring about negotiation with the Province over the area in question. Our goal was not to negotiate for a part of the proceeds from exploration, but to challenge the right of the Province to issue mineral claims and permits on lands that were covered under a comprehensive claim and which had never been surrendered or sold to the Crown. This challenge was part of our four-pronged approach to deal with the issue. If the Province did not have the right to issue permits or register claims filed by FVC, then it would be forced to pull the permits and revoke the claims.

The court process continued without our presence. FVC's contempt motion was heard by Justice Thompson, who ordered that the contempt charges be stayed for a period while the parties entered into mediated talks. We were given an opportunity to participate in this even though we had withdrawn from the court process. We agreed to the mediated talks in October of 2007, as winter was approaching and there seemed little likelihood of the issue being resolved through the continued occupation of the site. The mediated talks began in November and were flawed from the beginning as Mining and Northern Development refused to take any responsibility for the mess it had created by issuing the mineral permits and registering FVC's claims on lands under a comprehensive claim. The first seven weeks were spent arguing about the agenda for consultation. The position of the Province was that the staking had already occurred and nothing could be done about it. It wanted the agenda to begin from that point in time. The Province only planned to discuss where drilling would occur and how many holes there would be. Our position, on the other hand, was that consultation needed to go back and start from the point when no staking had occurred. We did not consider anything else to be meaningful consultation.

On the eighth week, the Province agreed to the creation of an agenda that would include an examination of the validity of staking. When we returned the next week to negotiate dates for the consultation to begin, we were shocked to discover that the Province had backed out of its agreement and relayed that consultation was dependent on us agreeing to pre-existing conditions. The precondition for our participation was our agreement that FVC could drill and continue its exploration work during the consultation process. We declined and walked away from the process. The failure of the parties to reach an agreement resulted in the resurrection of contempt charges, which had been stayed during the mediated talks.

When the case went back before the court, it was heard by Justice Douglas Cunningham, who had taken over the file from Justice Thompson. Justice Cunningham was angry that we had withdrawn from the court and assumed, with FVC's legal pleading, that our withdrawal was done to articulate a position of contempt for the Ontario court system, which was not the case at all. We felt at the time that the Ontario court system was incapable of providing a solution that protected our homeland from irresponsible development. We felt that our best chance of finding a resolution was to pursue a political solution. Our withdrawal from the court system had consequences because we were not allowed to file a defence against the contempt charges, which led to undefended judgments being issued against us by Justice Cunningham. This included the contempt charges and the $77 million lawsuit. FVC came away from this hearing thinking that it had won the day.

When we met in court again to deal with the sentencing for the contempt charges in February of 2008, we were allowed to provide evidence to the court in the form of testimony by our negotiator Robert Lovelace, who offered our collective reasoning for disobeying the court order to leave the site. Lovelace stated to the court that in many instances, Ontario law has harmonized itself with Algonquin law. In this particular case, however, Lovelace explained to the court that Ontario law violated Algonquin law in two important ways. First of all, Ontario law allowed the issuance of permits to a third party interest to conduct work that had been banned under Algonquin law. Secondly, Ontario law criminalized Algonquin citizens for protecting the lands and waterscapes within our homeland, which is required under the Guiding Principles, which articulate Algonquin law.[12]

In spite of the fact that Lovelace testified for two days and provided intricate details of Algonquin philosophy, values, and law, Justice Cunningham made the decision in his ruling to criminalize Ardoch Algonquin leadership and to portray us as criminals who deserved the most severe punishment for our defiance of Ontario law. Our positions as educators was mentioned in the ruling and the suggestion was made that were we to remain unpunished for our defiance of Ontario law we would incite insurrection in both Aboriginal and non-Aboriginal peoples in Ontario. Six-month jail sentences were handed down as well as fines for $25,000, $15,000 and $10,000. Robert Lovelace spent over one hundred days in jail before he was released in the successful decision of the Appeal Court on May 28, 2008. The Appeal panel stated that the sentencing was too harsh for people involved in a civil contempt process, which originated in an Aboriginal protest context. The Panel of Justices offered that no legal measures whatsoever should have taken place prior to the Province exhausting all possible attempts at meaningful consultation.

The Appeal Court also issued a decision about cost submissions related to the appeal in our favour, which ordered FVC to pay $40,000 and the Province of Ontario to pay $10,000. Ontario's portion was the result of its failure to deal with the issue prior to the court process. Three days later, Justice Cunningham released his decision on the costs related to the injunction itself, awarding $109,000 against us to Frontenac Ventures. We are presently appealing this decision and are awaiting a date for this in the Divisional Court. Unfortunately it appears that Justice Cunningham sits on the Divisional Court. At this point we are unsure of whether or not he will be recusing himself from what is obviously a conflict of interest. FVC has also filed leave to get permission to appeal the decision that freed Robert Lovelace from jail and stayed the fines. If they are successful, the case will go before the Supreme Court.

To date, Frontenac Ventures continues to conduct exploration work on our community lands without our permission and without meaningful consultation ever having taken place. Even our attempts to have them charged by the Ministry of Natural Resources (MNR) for infractions during the exploration process have failed. No one from the MNR would step forward to testify to the infractions in spite of the fact that we provided MNR officials with ample evidence of the damage done to wetlands by the construction of roads into the interior. Whole areas have also been clear-cut, eliminating the habitat of various animals, and a family trapline has been damaged as a result of this exploration work. Ardoch Algonquin people have been devastated by this project and the mining phase has not yet happened.

Finally, during this entire process, we have never been provided with scientific or other scholarly evidence from the Province that uranium exploration is safe. Nor has there been any type of environmental assessment to analyze the project for potential hazards to the region. There are several watersheds in the area that flow up into the Ottawa River and down into the St. Lawrence River. Yet, nothing has been done to reveal possible dangers to the watershed that provides sustenance to millions of people in the region. The allied resistance of Algonquin and private landowners has repeatedly asked for scientific or scholarly evidence that this project is safe, and we have continually been told to accept the word of Frontenac Ventures and Mining and Northern Development. We do not accept their word for this and have, out of necessity, begun a process of conducting our own research. The first stage of that research is presented in the pages to follow. The next stage will be conducted following exploration activities if we are unsuccessful in stopping the drilling process from occurring. It is our hope that it will provide a useful tool for other communities to assist them in their own struggle with the resource extraction industry.

STRUGGLING WITH URANIUM: THE IMPACTS OF EXPLORATION AND MINING

THE ONTARIO MINISTRY OF MINING AND NORTHERN Development (MMND), Aboriginal Affairs, and exploration company Frontenac Ventures would like Algonquin people and our neighbours to believe that there are no impacts from uranium exploration that we need be concerned about. When we voice our apprehension publicly as allies about this project, our collective anxiety is dismissed by Provincial officials as immaterial. What's more, in spite of the fact that uranium exploration is opposed by a large number of people in eastern Ontario, Ardoch Algonquin people continue to be portrayed as radical dissidents who are incapable of negotiating a solution to the impasse created around this project. We are expected to put the futures of multiple generations of our descendants in the hands of the mining industry simply because they say they have experts on their payroll who are prepared to inform us that uranium exploration is safe and harmless. Further, we are expected to trust these people even though their services are being paid for by the McGuinty government, which has openly declared that uranium exploration and mining is in the best interest of all Ontarians. As the community whose lands have been appropriated for this project, Ardoch Algonquin First Nation finds the attitude and actions of the Province to be extremely disrespectful and not in keeping with the *Honour of the Crown*, something Ontario is duty bound to uphold under the Proclamation of 1763.[13]

We find this current lack of respect for our cultural and spiritual rights to exist within our homeland to be very similar to the attitude held by nineteenth-century land speculator Philemon Wright, who was the first English squatter to appropriate land in Algonquin territory.[14] Wright was discovered by a group of Algonquin people clearing away a stand of maple trees in 1803. When he was confronted about this situation he lied and said he had documents from the English Colonial Office which gave him control of Algonquin land.[15] In reality he was a land speculator from Massachusetts who had come to secure land for a group of settlers. Wright acquired title to the land by agreeing to build the colonization roads into Algonquin territory for the English.

The Algonquin party he encountered asked him how he had gained ownership when they had not sold or traded their land to the English. Nor had they agreed that anyone could come onto their land and cut down their trees and chase away the game upon which they depended for their livelihoods.[16] Wright told them that he had a deed that gave him permission to be there and that they had to accept his word.[17] They replied that they would have been happy to have him leave their land and go back where he had come from, as they did not want their sugar bush disturbed and were concerned that his actions on their land would make it impossible for them to survive.[18] Wright told them that they had to be sensible, that the Colonial Office had given him the right to cut down all the trees he wanted. He also informed them that what he was doing would help them and that once he cleared the land and altered the region their lives would improve, as they would no longer need to live by the chase.[19]

Thirty-eight years later, Algonquin leader Shawanepenesi petitioned the English Crown for relief from the destruction of his community lands, saying that loggers were burning down the forest.[20] He reported that their village and all of the game upon which they depended had been smothered in thick black smoke from fires burning throughout the region.[21] Shawanepenesi, like the Algonquin people from his parents' generation, were told that what was happening was for the benefit of all and that he would receive no help from the Colonial Office. What's more, he was told that if he interfered with the loggers he and his community would suffer the loss of their annual presents.[22] This colonial process continued year after year and decade after decade with Algonquin people continually oppressed and forced to retreat into marshes and wetlands that were deemed unliveable wastelands by the colonial government and settlers of the day.

Ardoch Algonquin First Nation is a community of related Algonquin families whose members are the descendants of Shawanepenesi's community. Ardoch Algonquin families have historically occupied and related within the lands and waterscapes in the region of the Tay, Mississippi, and Rideau watersheds, all of which connect to the Ottawa and St. Lawrence Rivers through ground and surface water. The marshes and wetlands that became the permanent home of Algonquin communities such as Ardoch were always the traditional hunting, trapping, and harvesting areas of extended Algonquin families. The need for our ancestors to locate themselves permanently in the region was a direct result of colonial actions on the part of the English who appropriated the best lands within the Algonquin homeland for themselves.

There is a long history of interaction with our homeland that goes back for thousands of years. There are also numerous sacred and cultural sites within our homeland that have particular importance for us as they were left to us by our ancestors and other spiritual beings in Creation. Our ancestors passed on their knowledge and interactions with those places so that we would always have the ability to maintain our relationships and responsibilities within our homeland. The extended families that make up Ardoch Algonquin First Nation have struggled over the past two hundred years to hold on to our community lands and have begun to shake off the colonial oppression and assimilation that has kept us quietly contained and controlled. We have managed to find our voice once again and have begun to rebuild our relationships with the Natural World. This is difficult given the destruction that has occurred over the past two hundred years, but we have found ourselves again and have reinstituted our own laws and traditions within our homeland and our lives. We understand and recognize the fact that our identity as people and our autonomy within our homeland is dependent upon our ability to maintain our relationships and responsibilities.

As the above history illustrates, we cannot risk the future of our children and grandchildren and the health and vitality of the Natural World on the word of experts who work for a colonial government that continues to oppress us and deny our cultural and spiritual right to exist within our homeland. Nothing has changed in two hundred years except the cloak that surrounds the weapons that are used against us to achieve our oppression and domination. We are constantly told that we must accept these changes and the development happening on our lands because the government is acting in the best interest of everyone in Ontario. We have no choice but to object to this position, because we have no historical evidence that the government has ever acted in our best interests even though we are the original people of this homeland. What's more, we have no evidence that this project is in the best interests of our neighbours since numerous municipalities and counties have declared resolutions asking the Province to put in place a moratorium on uranium exploration and mining in eastern Ontario.[23]The Province has refused to act on these motions for a moratorium.

The Province prefers to move ahead pretending that there is no opposition. The Province should therefore not be surprised that we do not trust them to have the best interest of the Natural World and human beings in mind when making these decisions. They refuse to even consider the mountain of opposition that exists against this project. Algonquin people have decades of experience in the environmental and social consequences of resource exploitation within our homeland. Can the Province honestly blame us for being doubtful of their sincerity when even the premier himself supports uranium exploration and mining? We hardly think so and neither do our neighbours who are also worried about the impact this project will have on adjoining private lands.

We are dealing with lands and waterscapes that have already been devastated by two hundred years of continual domination. The land in question has been exploited previously, including the drilling of uranium core samples over the past forty years. The majority of those holes were never filled or capped and have more than likely contributed to the high uranium concentrations in some of the lakes and streams in the area. The Robertsville site contains evidence of another mining operation that occurred within the past thirty years, which has no doubt also added to the burden that the land and waterscapes have to carry.

Prior to the Robertsville development, the area had suffered from devastating logging practices that destroyed all of the old growth forest in the area. There are oral traditional stories about that time that tell us that those trees were as large as many of the redwood trees in California today.[24] This is substantiated by English documents that say that it took eight to ten men to cut down some of these trees. Unfortunately, those trees have disappeared as a result of irresponsible logging practices used in the nineteenth century. The impacts of resource exploitation are cumulative and increase the likelihood of ecological unbalance that can affect all aspects of Creation.

All of these activities and the thousands of others in between have disturbed and altered the land over the past two hundred years, thereby limiting the ability of Algonquin people to use the land. Those actions in our homeland have had multigenerational impacts with respect to social, emotional, physical, and spiritual well-being that are still being felt among our people today. These activities have also had multigenerational impacts on the settlers who established themselves here in our homeland. In a very similar way, our neighbours have begun to decolonize themselves and to also contemplate their relationships and responsibilities within the Natural World. We welcome this direction and respect it as a sign that it is possible to achieve existence within our homeland wherein the rights of the Natural World and those of human beings are considered together when contemplating resource development and exploitation.

The need to maintain that balance is the purpose for this study, to remind ourselves that we must consider carefully all of the ways that our actions can impact the land and the waterscapes upon which we all depend for survival. We need to consider the past, the present, and the future so that we make the best decisions for ourselves and for the Natural World. While we are certainly committed to pursuing the lessons available to us from the past, we have no interest in processes that would have us separate our lands and waterscapes into specifically identified sites or sections with varying degrees of "Algonquin value." Such an exercise is contrary to Algonquin law and epistemology (world view). Our community lands and those of our neighbours need to be seen and contemplated in their entirety as connected aspects of Creation that have breath and spirituality. This position provides guidance for participation in discussions related to resource exploitation and development that will allow all aspects of the Natural World to be considered together on an equal footing with the purported rights of human beings.

We must therefore consider the prospect of uranium exploration and mining in that context. Given that much of our land has already been altered over the past two hundred years, we need to examine the proposed exploration project in light of our need to maintain the land and waterscapes as they now exist with no further environmental damage or devastation. The land has already suffered from intense human activity, and therefore needs time to heal and regain balance. We must also consider the social, cultural, and spiritual losses Algonquin people have suffered as a result of previous actions on our lands, because that action has resulted in negative consequences related to our physical health and well-being. Since the health and vitality of Indigenous peoples is connected to the health and well-being of the Natural World, it makes sense that Indigenous peoples would value their relationships with their homeland and would seek to protect it in its entirety and nurture those relationships.

THE IMPACTS OF URANIUM EXPLORATION AND MINING

AS STATED IN THE INTRODUCTION, MMND, FRONTENAC Ventures, and Aboriginal Affairs would have us believe that there are no impacts to uranium exploration within our community lands or the private lands of our neighbours. This is an incorrect assessment, as there are no actions taken by human beings that do not have an impact on the Natural World. The very act of eating, of heating our homes and all the other necessities of life require that we use aspects of the Natural World for our survival. Everything has an impact! It is how we approach those aspects of Creation and how we place ourselves within the Natural World that determines the impact we will have as human beings within the land and waterscapes in our homeland. We are not children, even though we continue to be treated that way by the provincial government. We are capable of doing our own research and making our own decisions about development projects on our lands. It is important to remember that this conflict emerged in the first place because Algonquin people understood their relationship to the Natural World and the responsibilities we have to protect it. We stood up against this because we objected to the idea that permits would be issued and claims would be registered by the Province for an exploration project on lands where no notification or consultation had ever taken place with the affected communities (Algonquin and otherwise). Ardoch Algonquin First Nation and our neighbours questioned a process that would allow this to happen with no care for the health or well-being of our communities and the land we depend upon.

Understanding the relationship that human beings are required to have with all aspects of Creation, we stood up with our allies in opposition to a project where the rights of the mining industry were promoted above the Natural World and the human rights of Aboriginal peoples and local landowners. That is still our position today, a year later. We are still opposed to this project and do not support uranium exploration within our homeland. Consultation has never happened, and Frontenac Ventures continues to pursue exploration work on our community lands. To this day we have been provided with no absolute proof in writing that uranium exploration is safe nor have we been provided with a guarantee from the Province that Frontenac Ventures' work will not impact our lands and waterscapes and therefore our lives and those of our neighbours.

What we do know through our own research is that academic and scientific evidence exists that provides undeniable proof that there have been impacts from uranium exploration in locations around the world. There are four primary impacts associated with uranium exploration that are described in the literature that need to be considered. For Algonquin people, these impacts are important because they directly violate Algonquin law and our responsibility to protect the Natural World. The first impact is the contamination of ground and surface water that can happen as a result of the drilling of core samples. The Navaho Nation is a good example of a place where such contamination took place. The Navaho Nation, which has been mined for over forty years, has over a thousand open-pit shafts that were never filled or remediated.[25] The Navaho people have suffered contamination to both their ground and surface water from exploration drilling and mining.[26] This is no surprise as "drilling has the potential to impact the environment in a variety of ways."[27] Likewise, drilling activities in the Black Hills in North Dakota resulted in several wells in the exploration area dropping substantially.[28]

Subsequently, they determined that the drop was caused by the mixing of water between the drill site, adjacent aquifer, and people's drinking water.[29] Just recently, the Grand Canyon in the United States became the object of a bitter conflict between exploration companies, state governments, environmentalists, and Indigenous peoples. The federal government stepped in and put an end to the dispute by outlawing exploration and drilling within the Grand Canyon. [30]

Contamination of ground and surface water not only affects regional ecosystems, it also violates Algonquin law, which holds waters to be sacred. Anishinaabe Elders who participated in a workshop on water reiterated its importance to us as human beings:

> The Elders stated that *sah-kemah-wapoye*, which translates as 'spiritual water,' refers to the living water. In the spiritual sense, water is an entity that can be damaged by humans who are manipulators. We can make things happen, we can move things, we can throw things and we can break things. They characterized water as more passive, in that it will accept things done to its essence. Water will accept these changes and will undergo the same changes. It will mirror the climate or mood that we, as human beings, are in. It becomes the quality in which we shape it. We humans, in turn, are affected by the changing quality of water, and say that the water is 'hurting us.' It is only giving us what we have asked of it.[31]

Clearly, from the perspective of the Elders participating in the workshop, human beings have particular responsibilities to water and have to be careful of our decisions, as water will change and mirror our actions.

Elders shared in the workshop that water is sacred because it is life. They cautioned that we must use care because "water is one of the bases for life, second to air. If we eliminate water, humans will not survive."[32] They further offered that "If we abuse water we will lose it; it goes back into the ground like some medicines that are abused. Mother Earth will take away the water and won't return it until we learn to use it properly."[33] Given the sacred nature of water and the protection it has within the legal systems of Indigenous peoples, it should be no surprise to the Province and Frontenac Ventures that we have similar provisions within Algonquin law, which forbid contamination of ground and surface water.

The second impact that has been associated with exploration and drilling is the release of radon gas into the immediate area around the drill holes. "Breathing or ingestion of abnormal levels of the radioactive gas radon, derived from natural sources such as rocks, has been considered as carcinogenic and [causing] kidney-related diseases."[34] Gordon Edwards, who is the president for the Canadian Coalition for Nuclear Responsibility, warns that concerns over health are warranted because radon gas is dangerous. He argues that huge amounts of it can be "released into the air, and dissolved in surface waters [from exploration drilling and mining]." What's more, the "US Surgeon General has determined that radon is the second leading cause of lung cancer after cigarette smoking; tens of thousands of Americans die every year from exposure to radon gas."[35] Jonathan Samet and George Eradze further relate in their study that radon is a well-established human carcinogen for which extensive data is available.[36] "Radon is a noble and inert gas resulting from the decay of naturally occurring uranium-238. With a half-life of over three days, radon has time to diffuse through rock and soil after it forms and before undergoing further decay into its particulate progeny."[37] Unlike radon, Samet and Eradze contend that the progeny of radon are "solid and form into small molecular clusters or attach to aerosols in the air after their formation. The inhaled particulate progeny may be deposited in the lung on the respiratory epithelium; radon by contrast is largely exhaled, although some radon is absorbed through the lung."[38]

While we are told there are no impacts from exploration, research such as this clearly reveals that there is a danger of radon gas being released through drill holes and fractures in rock formations caused by drilling during the exploration phase. This is of particular importance, as radon and its progeny pose a serious health threat to people and animals in the vicinity. Samet and Eradze argue that sufficient evidence exists to directly link the development of lung cancer to alpha particle emissions from radon progeny.[39] Radon also has the potential to be carried far from drilling sites by wind and foul weather.[40] Despite its relatively short-term half-life of 3.8 days, radon presents a long-term hazard for people and the environment.[41]

Along with the danger of ground and surface water contamination and the hazards associated with the release of radon gas and its progeny, uranium exploration also destroys the habitat of our relatives in the Natural World. Exploration often includes the removal of overburden and the construction of roads into the areas where drilling will take place. For instance, advanced exploration may use trenching and drilling, and access roads, airstrips, and exploration camps may eventually be constructed, with increasing potential for impacts on the environment:[42]

> One impact created by these roads is erosion, which may cause disruption to habitats of fish. A larger impact comes from opening up access to potentially sensitive wilderness areas, such as alpine meadows or caribou calving grounds. Hunting, wilderness tourism, guides and outfitters, and other non-industrial users may cause significant environmental impacts, from noise to harassment of wildlife.[43]

As is stated here, habitat loss can occur as a result of uranium and other mineral exploration. This is significant to communities who depend upon their lands for cultural, spiritual, and subsistence needs.

What is further distressing to Indigenous communities are the trails, roads, buildings, and other materials built during exploration projects that are left behind after mineral exploration activities have concluded. This disrespect for the land is evident in Nunavik, where there are 595 abandoned uranium exploration sites with a variety of buildings and materials, all left behind by exploration companies. In the 1990s, Inuit hunters informed the authorities of the presence of oil and chemical products on specific sites and mentioned the progressive deterioration of storage conditions, but very little was ever done about it:[44]

> Over the years, Inuit and Naskapi hunters have reported a wide range of equipment on the sites, including cans of food, modular laboratories, heavy machinery and equipment for the storage and transportation of oil products. In the mid-1990s, the Inuit population became increasingly concerned about the presence of abandoned chemical and oil wastes: storage conditions had deteriorated and the containers were no longer securely sealed. Hunters noticed various serious environmental problems on a regular basis such as the death of foxes resulting from the consumption of various rotting food products

and highly toxic chemical products near streams, lakes and rivers.[45]

Clearly, there is a history of negligence on the part of exploration companies, which appear to have little regard for the Natural World or the people who live in the region. These companies did not submit their exploration work to environmental assessments prior to beginning work, and it is obvious from the enormous number of abandoned sites that they did not bother to perform remediation either. Given this past record, how could anyone fault Indigenous peoples for the fear they carry about this type of development on their lands or those of private landowners?

Uranium exploration is also an issue for Akaitcho Dene and the Athabasca Denesuline peoples in the Arctic who face possible impacts from four uranium exploration projects in the Upper Thelon Region. This is a highly sensitive issue as this is the traditional migration area for certain herds of caribou. These communities are deeply concerned about the cumulative effects of repeated disturbance on habitat from multiple exploration activities within the same small area, especially given the fact there has been previous exploration work in the area.[46] Concern is also growing about the contamination that may accumulate in caribou as they migrate through their seasonal ranges over a number of years.[47] What befalls the caribou will befall the Dene because they depend a great deal on healthy caribou for their survival.

West Arnhem Land in Australia has also undergone various impacts as a result of uranium exploration. They have had to cope with the "spread of weeds and feral animals, and erosion"[48] as a result of various exploration projects. These communities have also had the additional burden of reduced traditional access to lands and have identified this as an issue of ongoing concern.[49] They argue that "Aboriginal concerns about uranium need to be seen within the context of past actions by the mining industry and the government, the bulk of which have ignored the basic land and human rights of Indigenous peoples."[50]

In the U.S., habitat impacts have also been reported from uranium exploration in the Alcova Reservoir and Muddy Mountain area, which lies just outside Casper, Wyoming. A significant amount of uranium exploration occurred in the drainage in the 1970s.[51] This exploration and subsequent lease maintenance resulted in the creation of numerous roads and trails, some of which have contributed to accelerated erosion. [52] The actions of those resource exploration companies led to degradation of the landscape, which, of course, has implications for people, animals, and amphibians in the region.

Wyoming residents have a lot to be concerned about with new uranium projects because of this and other previous exploration activity as it leads to cumulative effects that multiply over time. In addition to the work done in the Alcova Reservoir and Muddy Lake area, there are over 350 abandoned uranium exploration sites in the Bighorn Canyon National Recreation Area in nearby Montana. These sites "vary in size from small D6 caterpillar scoops to large complex excavations of one-half acre."[53] These sites were created between 1956 and 1960, and still show the two-track mining roads used for access.[54] The National Park Service began a reclamation project in 2003 to implement measures to restore the site because of the eyesore it presented to the public. Park officials felt that "the multiple disturbances from mineral exploration were aesthetically distracting and could result in an inappropriate visitor interpretation of the landscape."[55] In fact, some of the lands were so damaged that fragmentation of various habitats has occurred.[56] What's more, an environmental assessment completed by the Park Service in 2003 revealed that the search for uranium in the late 1950s had left gouges deep enough to change the topographic features of the park in three areas (Devil Canyon Overlook, Barry's Island and south of Layout Creek Canyon).[57] While the Park Service was required to submit its reclamation project to an environmental assessment, the companies that caused the damages and impacts on the landscape through uranium exploration had no such requirement.

In Wyoming and Montana, like many other place in the world, much of the damage inflicted from these projects happened because uranium exploration projects are not often scrutinized through an environmental assessment process prior to the issuance of permits and leases by governments. Without this necessary oversight, the exploration industry is left to regulate itself, which does not appear to be as important to companies as economic considerations. Self-regulation also creates a conflict of interest, as many of the lands exploited by mineral exploration companies are Indigenous or public lands. It is no wonder that Indigenous nations are deeply concerned about the impacts of uranium exploration; there is a history of numerous companies conducting exploration campaigns without showing any concern for the impact this type of activity has on the environment.[58]

Indigenous communities and private landowners have no faith in the ability of companies to self-regulate, especially given the history and track record of the industry. Part of the mistrust that develops around these projects has to do with the culture maintained by the mineral exploration industry. In a report to the Prospectors and Developers Association of Canada, Ian Thompson and Susan Joyce remarked that "the culture of mineral exploration does not encourage good community relations."[59] The authors argue that often conflicts develop between

Indigenous peoples and exploration companies because the culture espoused by exploration personnel is "arrogant, insensitive and very threatening."[60] While this source did not deal specifically with private landowners, one could apply this scenario to them as well as they no doubt also experience the effects of this "culture."

The prevailing attitude among exploration companies appears to be one of avoiding or limiting interaction with local communities unless there is some compelling reason to do more.[61] This attitude arises from the fact that mineral exploration is highly competitive and secretive.[62] They do not want the word to get out about a project until they know that their interests have been firmly protected. Junior companies, in fact, are strongly oriented to the venture capital markets and thus focused almost exclusively on the technical aspects of a project.[63] Community relations are, at best, a secondary consideration according to Thompson and Joyce.[64]

Given the secrecy of the exploration industry, many Indigenous communities and private landowners find out after the fact that their lands have been appropriated under antiquated mining laws and turned over for resource exploitation. Having had no prior notification or consultation, communities are forced into a position of reaction against something that, under the circumstances, cannot help but be interpreted negatively. Indigenous communities and private landowners alike feel betrayed, albeit for different reasons. Private landowners feel betrayed because for the most part they have been socialized to believe that they live in a democratic society in which the individual rights of private landowners are of paramount importance. Most Indigenous peoples, on the other hand, are fully cognizant of the fact that their rights have never mattered to anyone. What Indigenous peoples get angry about are the constant betrayals and lack of respect for the right to exist as autonomous peoples within our own homelands.

Betrayal and dismay are often followed by feelings of outrage and anger when communities discover that there is no process wherein these projects are properly assessed. Without an environmental assessment process, there is no need for these companies to deal with Indigenous communities or private landowners as stakeholders. Supported by environmental legislation, companies are left to perpetuate the culture of exploration, as described by Thompson and Joyce. They secretly acquire permits and claims and quietly go about staking Indigenous, public and private lands, secure in the knowledge that they are beholden only to their stockholders. By the time many Indigenous communities and private landowners discover their lands have been staked it is too late to stop some of the impacts that happen as a result of exploration activities. Trails and roads are built, trees are cleared, and overburden removed. In some cases, all that is left for the company to do is bring in a drill.

Thus the fear and terror experienced by Indigenous peoples over proposed uranium exploration projects on their lands is not only understandable, it is entirely justified. It is fear that has been created out of the experiences other communities have had with exploration companies. It is also a fear that emerges out of the culture of exploration wherein communities are alienated from the approval process. The fear is widespread and shared, for instance, by the Dene, who do not put a lot of faith in the ability of exploration companies to regulate themselves:

> We do not have confidence in the use of *Saskatchewan Best Practices Guidelines* to effectively address our concerns about uranium, and similarly do not believe that other mitigation measures imposed as conditions of a land use permit would be sufficient to prevent adverse impacts from occurring.[65]

The Lutsel K'e Dene First Nation (LKDFN) have consistently and repeatedly voiced their opposition to mineral exploration in the Thelon Basin on the grounds that the entire Thelon watershed is of vital importance on many scales, and the upper Thelon in particular has distinct and unique environmental, cultural, spiritual, and heritage values which would be at significant risk should mineral exploration and/or mining be allowed to proceed.[66]

The worries and fears are the same across Indian country as well as among private landowners in North America. The mining industry and many government officials want us to back down and accept their word that uranium exploration is safe when there are tangible examples where these projects have contaminated ground and surface water; exposed human, plant, aquatic, and animal life to increased doses of radon and its progeny, and caused habitat fragmentation and destruction. The evidence presented thus far substantiates this point. Not only is it possible for accidents to happen during uranium exploration, it is also the case that some exploration companies are negligent in their work and in their approach to dealing with Indigenous communities and private landowners.

The fear and anger is understandable, and is the place where we find ourselves right now as an Algonquin community. This fear and anger is the result of actions on the part of the Province and Frontenac Ventures. Our lands have already been affected and we have not even been consulted yet. Many of the impacts associated with uranium exploration have already happened to our community lands and the diamond drilling has not even begun. Frontenac Ventures built trails and roads into the interior of our lands, so that they would have easy access to bring in a diamond drill. In the process, they dumped gravel and road materials into wetlands that connect with other ground and surface water in the area. They have also destroyed habitat, including part of a trapline that one of our extended families depends upon for subsistence.

This particular wetland is home to beavers and other animals, such as amphibians. It is also on the path of migrating birds. The actions of Frontenac Ventures have resulted in impacts to the land and the waterscapes, which jeopardizes our ability to maintain balance in that wetland. Between October 2007 and July 2008, Frontenac Ventures built additional trails and roads, cleared away forest, removed overburden, and scraped away rock facing in preparation for drilling core samples, which left the porous rock exposed. This activity is absolutely against the wishes of Ardoch Algonquin First Nation and some of our neighbours who also oppose the project. Numerous complaints have been made to the Ministry of Natural Resources (MNR), and after continual onslaughts of emails, phone calls, and letters, infractions have been laid against the company and some of its associates.

The infractions, though, cover only the initial damages done prior to June 2007. No charges have been laid against the company for the damages they have inflicted since we left the site in October 2007. The Ministry has refused to impose a stop work order on the company as we have requested. Nor has it ordered an environmental assessment on the remainder of Frontenac Ventures' work plan even though the MNR was provided with absolute proof that additional damages have taken place since we left the Robertsville site last October. Thus Frontenac Ventures is allowed to remain outside the normal environmental assessment process that all private landowners in Ontario are required to follow. The frustration felt by communities faced with this situation is enormous. All the mechanisms normally used to combat the problem have proved incapable of stopping this encroachment.

In this case, the government of Ontario remains complacent about our oppression and continues to support the economic rights of Frontenac Ventures over the health and vitality of our people and the land and waterscapes we depend upon for our survival. This continues to happen in spite of the Court of Appeal decision, which articulates the need for meaningful consultation prior to the approval of resource exploitation. While Ardoch Algonquin First Nation remains committed to participating in consultation with no preconceived notions about the outcome, there have been no offers of consultation from the government that do not include preconditions to allow drilling. We have been informed that all of Frontenac Ventures' rights are to be upheld in any consultation process while there is no mention of our social, cultural, and spiritual rights within our community lands.

Imposed preconditions leave no opportunity for discussions that get to the bottom of Indigenous community concerns because the agenda is crafted in advance by government officials and exploration companies. In most instances, staking has already occurred and governments discourage discussions that question the validity of staking, land use permits, and claims on Indigenous and private lands. Communities are expected to participate from the point that those things have already happened and the most that can be achieved through consultation is mitigation of impacts on sites of social or spiritual significance. Mitigation is useless in those cases where the damage to wetlands and forests from exploration activities has already happened. In situations such as ours with Frontenac Ventures, the only things left to mitigate, according to the Province, are the number of drill holes and where they will be located.

Mineral exploration is not just a culture that promotes arrogance, insensitivity, and intimidation; it is also a culture that absolutely promotes the oppression of Indigenous peoples. What's more, the same culture of disrespect exists within the Ministries of Mines and Northern Development, Natural Resources, and Aboriginal Affairs. These ministries support and promote the economic rights of resource extraction companies over the rights of Indigenous peoples and private landowners. This is evident in the sham consultation process that Ontario promotes wherein all the major decisions are made with no input from the affected Indigenous communities. Communities that refuse to participate in this discriminatory process are left to hang in the wind, their environmental, health, cultural, and spiritual concerns buried under layers of whitewash.

Environmental concerns are not the only impacts that Indigenous communities must consider when threatened with resource exploitation. Development also has social, cultural, and spiritual consequences that can reach across generations. Even when an assessment is applied to development projects, the social, cultural, and spiritual impacts of concern to Indigenous peoples are not given the same priority as environmental concerns. One avenue available to communities today that provides a method of analyzing possible social impacts should an assessment be ordered can be found within the dimensions of *Social Impact Assessment (SIA).* SIA is a social research tool developed by sociologists in the 1970s. It allows communities to assess or estimate, in advance, the social consequences that are likely to follow from specific policy actions or project developments.[67] Burdge and Vanclay report that within SIA, social impacts include all social and cultural consequences to human populations of any public or private actions that alter the ways in which people live, work, play, relate to one another, organize to meet their needs, and generally cope as members of society.[68]

In general, an SIA process provides direction for the following:

- understanding, managing, and controlling change;
- predicting probable impacts from change strategies or development projects that are to be implemented;
- identifying, developing, and implementing mitigation strategies in order to minimize potential social impacts (that is, identified social impacts that would occur if no mitigation strategies were implemented);
- developing and implementing monitoring programs to identify unanticipated social impacts that may develop as a result of the social change;
- developing and implementing mitigation mechanisms to deal with unexpected impacts as they develop; and finally,
- evaluating social impacts caused by earlier developments, projects, technological change, specific technology, and government policy.[69]

The authors believe that SIA should be required of all public and private activities (projects, programs, policies) that are likely to affect social life.[70] Most Indigenous communities would agree that the social impacts of any development project should be assessed prior to the approval stage.

> This is important as millions of people lose their livelihoods as a result of mining: Mining is the root cause of numerous civil wars, dictatorships, foreign armed interventions, widespread human rights abuses, poisoning of people and environment, deforestation and forest degradation. It is true that humanity needs a certain amount of minerals to satisfy its basic needs and it is also equally true that over-consumption by one part of humanity is destroying the livelihoods and environments of the other humanity at the receiving end of mining. Mining is an activity that needs to be strictly controlled at all stages. Above all, people living in mining areas should have the capacity to make fully informed decisions on the permission to mine in the territories or decide on how to carry out the activity and ensure environmental conservation and social justice.[71]

Millions can testify to the high social costs that comes with mineral exploitation and the processes that accompany it, such as the appropriation of land belonging to the local communities, impacts on health, alteration of social relationships, destruction of forms of community subsistence and life, social disintegration, radical and abrupt changes in regional cultures, and the displacement of other present and/or future local economic activities.[72]

Women are also disproportionately affected by resource exploitation and development, including mineral exploration and mining. Ricardo Carrere reported on this pattern in 2004, arguing that women are often forced to carry the burden of such development.[73] The following provides a concrete example of the ways rural women in Indonesia are being impacted by mineral exploitation:

> Mrs. Satar had a field as large as 10 to 15 hectares on the community's traditional land. Upon this land, she could harvest enough produce for one year, in fact sometimes more. With the introduction of the mining into her community, she lost all but one hectare of her land to the mining company. Consequently, she had to buy approximately three sacks of rice per month at a cost of Rp39,000 per sack (price at January 1998). In addition, the mining company's operations polluted the river, which could no longer be used to meet household needs, and no longer produced fish. Previously, Satar had cooked fresh fish each day for her family. Now, as a result of the pollution, she has to buy salted fish. If there is enough money, she purchases 2 kilos of salted fish a month at Rp15,000 per kilo. To obtain bathing and drinking water, Satar must walk a long way to a water source that is not affected by the company's tailings. Satar's livelihood is further threatened by the loss of her two water buffalos, found dead at the edge of the contaminated river.[74]

With the onslaught of exploration and mining activities in the lands of Indigenous peoples, women find that their roles and responsibilities are enormous as they try to cope with the social consequences of resource exploitation.

Victoria Tauli-Corpuz, who was the director of Tebtebba Foundation in 1997, argued that resource exploitation led to "significant erosion or destruction of traditional values and customs which have been crucial in sustaining community, tribal, clan and family solidarity and unity."[75] Furthermore, she warned that the traditional roles women have had as food gatherers, water providers, caregivers, and nurturers have also been affected.[76] The increased burdens forced on women have led to more stress and tensions, and, for some, even mental illness.[77]

Along with environmental and social costs that accompany resource exploitation, communities must also contend with the various cultural impacts of mineral exploitation. Cultural impacts, according to Burdge and Vanclay, are those "changes to the norms, values, and beliefs of individuals that guide and rationalize their cognition of themselves and their society."[78] They recount that once local cultural life is affected by development, it is affected for good; therefore, it is important to prevent the majority of impacts before they actually happen.[79]

Ciaran O'Faircheallaigh identifies two dimensions of cultural heritage that must be considered when discussing resource development on Indigenous lands. The first dimension involves material manifestations of Aboriginal occupation in earlier periods of time.[80] This would include burial sites, middens created by discarded shells and other food debris, rock and cave paintings and scatters of stone tools. These manifestations could be up to 50,000 years old, or only a generation or two removed from the present.[81] The second may be lacking in material manifestations but contains places, sites, or areas of spiritual significance to living Indigenous peoples.[82] These sites of spiritual significance are often associated with Stories of Creation, when spiritual beings moved across the landscape and created not only the forms the land now takes, but also the law that governs people's interactions with the land and each other and the languages and ceremonies that constitute key elements of Indigenous cultures.[83]

O'Faircheallaigh contends that while outsiders may perceive marked differences between these dimensions of cultural heritage, Indigenous peoples do not. Within Indigenous homelands, certain sites remain the resting place of powerful creation spirits. Other sites may be important because they are harvesting grounds for key food or medicine plants. Some places are associated with initiation, mortuary, or other ceremonies, or because they were the location of important historical events.[84] What must be understood and recognized is that all of the land and waterscapes in Indigenous homelands contain knowledge and laws associated with those sites as a single entity that must be protected as a whole. Indigenous peoples see themselves as not only connected intimately to their homeland, but also to earlier generations who have used the land and to later generations who will use it in the future.[85]

The understanding that land and waterscapes must be seen as components of an interconnected homeland was also relayed by Chief Adeline Jonasson of the Lutsel K'e Dene First Nation in a letter to former Indian Affairs Minister Jim Prentice:

> The mining special interests have a profound misunderstanding of the intrinsic significance of the upper Thelon from the perspective of the Dene people, and indeed of many Canadians. The significance of the Thelon to our people as the center of our civilization, our history, our spirituality and our lifestyle cannot be overstated. To allow mineral exploitation to proceed in the region without careful planning and consideration is tantamount to allowing the desecration of our church, our grocery and our museum. Such sacrilege is nothing less than expediting the demise of our unique culture and way of life.[86]

Clearly, the cultural impacts of mineral exploration are of paramount importance to Dene people. Their homeland is the centre of their culture, language, and spirituality. They oppose uranium exploration because of the impacts it will have on their way of life.

Chief Jonasson is not alone in her determination to avoid the impacts associated with mineral exploration. The Hopi, Navaho, and Apache have also passed resolutions against uranium exploration in their own lands. They believe that drilling exploration wells throughout the watersheds of Mt. Taylor violates their religious freedom. They hold that "the Mt. Taylor region and source water originating there are sacred and indispensable to their traditional cultural practices."[87] The Pueblo Council, cognizant of the links between contamination of the watershed and the subsequent impacts on their cultural heritage, chose to pass a resolution banning uranium exploration within their homeland. Given the importance of their relationship with the Natural World, the Council believes that the cultural rights of tribes to lands traditionally used for subsistence and cultural activities from time immemorial should be protected under religious freedom and environmental justice principles.[88]

The Pueblo resolution points to an additional impact from mineral exploration that has not been discussed in any detail to this point. The Pueblo Council contends that impacts from uranium exploration constitute a violation of their religious freedom. Given the fact that similar feelings have been expressed about mineral exploration by numerous Indigenous communities around the world, it would be pertinent at this juncture to provide a bit of context for that discussion. The basis for this belief can be found within all Indigenous epistemologies or world views because they provide the foundation wherein people perceive the world, and the place of human beings in that world. Chief Jonasson stated this relationship succinctly when she relayed that the Thelon region is the center of Dene civilization, history, spirituality, and life.[89]

Tewa scholar Gregory Cajete conceives of this intimate relationship between Indigenous peoples and their homeland as one of "ensoulment."[90] Indigenous relationships, he contends, exist within a *spiritual ecology* that developed over thousands of years of relating in one's homeland:

> Through generations of living in America, Indian people have formed and have been informed by the land. Kinship with the land, its climate, soil, water, the mountains, lakes, forests, streams, plants, and animals have literally determined the expressions of an American Indian theology. As a result of this intimate relationship, the land has become an extension of Indian thought and being because, in the words of a Pueblo Elder, 'it is this place that holds our memories and the bones of our people…this is the place that made us.'[91]

Cajete believes that this ecological awareness has led to the development of a complex relationship between Indigenous peoples and the North American landscape that has existed for at least 30,000 years.[92]

In this context, Cajete informs us that the paradigm of thinking, acting, and working on the part of Indigenous peoples evolved because of and through an established relationship to nature. As such, the foundation, expression, and context of Indigenous education and perceptions of history were environmental. [93] What's more, the theology of nature reverberated throughout art, community, myth, and any other aspect of human or social expression. All were inspired and informed through an integrated and direct relationship to the Natural World.[94]

From Indigenous understandings of Creation and the responsibilities that emanate from Creation, social and political structures emerged within Indigenous societies that were distinct and oriented to place. These social and political structures emerged as a mechanism to enable Indigenous peoples to maintain their relationships and responsibilities within their homeland. Over time, natural laws developed around those responsibilities, which were then incorporated into Indigenous social and political structures. Those laws provided the guidance necessary for a collective historical consciousness to develop that was transferred from generation to generation through oral tradition, cultural documents,[95] and ceremonial practices.

As Cajete articulates, the spiritual ecology in which Indigenous peoples exist is spread throughout the entirety of a homeland and cannot be separated into separate values according to specific sites. To attempt such an exercise is futile, as sites are connected with each other across space and time. They are also part of the same ecological context into which Indigenous peoples were created.[96] It is the existence, in fact, within this spiritual ecology that brought about the need for social and political mechanisms to maintain balanced relationships with one's homeland. Education then within Indigenous epistemologies was not concerned with celebrating the accomplishments of human beings. Instead, it was focused on passing down the knowledge through ceremonial practice and teachings that would help to shape individuals into complete human beings who could live in balance within their homeland.[97] Presented in this way, one can begin to understand the complexity inherent in the relationships that Indigenous peoples maintain with their homeland.

As illustrated here, there are numerous considerations that have to be taken into account when dealing with the possibility of resource extractions on Indigenous lands. Those examinations must take place within Indigenous epistemological frameworks so the multifaceted nature of impacts to Indigenous social, cultural and spiritual dimensions can be properly assessed. Attempts on the part of governments to force Indigenous communities into consultation processes that have predetermined outcomes or to identify and rank sites with specific values that reside outside Indigenous cultural contexts is ill-advised because they do not address Indigenous concerns and will prove incapable of resolving the underlying issues that led to the conflict in the first place. Respect for the position of Indigenous peoples as the centre of the wampum belt must be acknowledged and respected if these conflicts are to be resolved.[98]

This is the cultural and spiritual context in which we speak as Ardoch Algonquin people against this particular resource development project on our community lands. As Anishinaabe people, we understand that we were moulded from the Earth and that we have the Creator's breath in our bodies. We acknowledge the fact the breath we are breathing now is the same breath used by our ancestors in the past.[99] We believe that we are part of the land and the land is part of us. As human beings we have no other place in the world that we can call our home. Because of that relationship we were given Original Instructions to care for the Natural World. Those Original Instructions were maintained through the principles of Pimaadiziwin and the Teachings of the Seven Grandfathers.[100]

Pimaadiziwin is the Algonquin term that expresses the spiritual ecology in which we exist as human beings. It is the epistemological theory that shapes our understandings of the world and our relationships with the Natural

World. Pimaadiziwin refers to the ability to live the "good life" with a "good heart and mind."[101] The Teachings of the Seven Grandfathers are the methodology we use to guide our attitude and behaviour with each other and the world around us. Those teachings include honesty, bravery, wisdom, truth, honour, love, and respect. Pimaadiziwin and the Seven Grandfather Teachings were integrated into each and every aspect of individual and collective existence.[102] Pimaadiziwin, then, shaped the behaviour of individuals in ways that prevented them from making choices or decisions that would jeopardize the health and vitality of the community.

BIBLIOGRAPHY

Adeline Jonasson, Chief. *North of 60 Mining News*, July 29, 2007; available from <www.petroleumnews.com/pntruncate/484739216.shtml> accessed August 1, 2008.

All Indian Pueblo Council. *Resolution 2007–12: Companion Resolution for the Protection of Mt. Taylor and all Sacred Sites and Cultural Properties Related to the Pueblos of Acoma and Laguna.* New Mexico: All Indian Pueblo Council, June 2007.

Ardoch Algonquin First Nation. *Principles for Development.*

Arizona Game and Fish Department. "Uranium Mining and Activities, Past and Present," *Update for the Arizona Game and Fish Department Commission.* Arizona: Arizona Game and Fish Department, May 2007.

Bate, C.A. Personal Communication, March 18, 2005.

Brugge, Doug, and Rob Goble. "The History of Uranium Mining and the Navaho People," *American Journal of Public Health* 92 no. 9 (2002).

Burdge, Rabel J., and Frank Vanclay. "Social Impact Assessment: A Contribution to the State of the Art Series," *Impact Assessment* 14 no. 1 (1996).

Cajete, Gregory. *A People's Ecology: Explorations in Sustainable Living,* Santa Fe: Clear Light Publishers, 1999.

Catholique, Iris. "Letter to Alan Ehrlich—Senior Environmental Assessment Officer," *Environmental Review.* Mackenzie Valley Environmental Impact Review Board (May 30, 2008).

Carrere, Ricardo. *Mining: Social and Environmental Impacts.* Montevideo, Uruguay: World Rainforest Movement, 2004.

Chinkhanmuan, Gualnam. "Mining Social and Environmental Impacts," *Asia Indigenous Peoples Pact Foundation* (April 30, 2008).

Department of Primary Industries. "Minerals & Petroleum," Department and Primary Industries Online; available from <www.dse.vic.gov.au/DPI/nrenmp.nsf/9e58661e880ba9e44a256c640023eb2e/61839ca05e2f0250ca2573b600806acb/$FILE/ATTPZDJN/2%20GFEM%20-%20Drillholes.pdf> accessed March 3, 2008.

Duhaime, Gerard, Nick Bernard, and Robert Comtois. "An Inventory of Abandoned Mining Exploration Sites in Nunavik, Canada," *The Canadian Geographer* 49 no. 3 (2005).

Edwards, Gordon. "Press Release in Support of Uranium Protest at Sharbot Lake," *Mining Watch Canada List Serve.* [homepage] available from <lists.miningwatch.ca/pipermail/news/2007-September/001555.html> accessed May 13, 2008.

Environment Canada. "Human Activities, Minerals Metals and Mining, Environmental Impacts of Mining Exploration," *Information Data Base: The State of Canada's Environment 1996,* Environment Canada Online [home page] available from <www.ec.gc.ca/soer-ree/English/SOER/1996report/Doc/1-7-4-7-6-2-1.cfm> accessed June 20, 2008.

Grinde, Donald A. Jr., and Bruce E. Johansen, eds. *Ecocide of Native America: Environmental Destruction of Indian Lands and Peoples.* Santa Fe: Clear Light Publishers, 1994.

Higginson, J.M. Memorandum. NAC RG 10, Vol. 186, pt 2, pg 108, 566 Reel C–11, 511.

Hurley, Mary C. *The Crown's Fiduciary Relationship with Aboriginal Peoples* Ottawa: Parliamentary Research Branch: Law and Government Division, 2000, pp. 1–11.

Jarvis, Colonel. Report. NAC RG 10, Vol. 186, pt 2, pg 108566I-108566K Reel C–11, 511.

Johansen, Bruce E. "The High Cost of Uranium in Navaho Land," *Akwesasne Notes* New Series 2 (April 1997).

Joyce, Susan A., and Magnus MacFarlane. "Social Impact Assessment in the Mining Industry: Current Situation and Future Directions," *Mining, Minerals, and Sustainable Development* 46 (2001).

Ka-Pishkawandemin Family Head's Council Meeting. Ardoch, Ontario, November 26, 2006.

Lavalley, Giselle. "Aboriginal Traditional Knowledge and Source Water Protection: First Nations' Views on Taking Care of Water," *A Report Prepared for the Chiefs of Ontario and Environment Canada* (March 2007).

Lovelace, Robert. "Testimony Before the Ontario Court," Kingston, Ontario, February 14, 2008. Lynden, Justice. Ipperwash Inquiry, <www.attorneygeneral.jus.gov.on.ca/inquiries/ipperwash/index.html>.

O'Faircheallaigh, Ciaran. "Negotiating Cultural Heritage? Aboriginal–Mining Company Agreements in Australia," *Development and Change* 39 no. 1 (2008).

Paci, C., and N. Villebrun. "Mining Denendeh: A Dene Nation Perspective on Community Health Impacts of Mining," *Pimatisiwin: A Journal of Aboriginal and Indigenous Community Health* 3 no. 1 (2003).

Reid, Christopher. "Legal Brief the Duty to Consult," prepared for the Uranium Defence website: <www.uraniumdefense.ca>.

Salem, H.M. *Uranium Ores and the Environmental Impact on Human Health Risks.* Cairo: Cairo University Press, 2000.

Samet, Jonathan, and George R. Eradze. "Radon and Lung Cancer Risk: Taking Stock at the Millennium," *Environmental Health Perspectives* 108 Supplement 4: Occupational and Environmental Lung Diseases (Augist, 2000).

Scott, Gary, and Mark Wakeham. "Uranium Exploration in West Arnhem Land: A Report for the Environment Centre Northern Territory and the Australian Conservation Foundation," *Australian Conservation Foundation* (2001).

Sherman, Paula. "Indawendiwin: Spiritual Ecology as the Foundation of Omàmìwinini Relations," A Dissertation Submitted to the Committee on Graduate Studies in Partial Fulfillment of the Requirements for the Degree of Doctorate of Philosophy in the Faculty of Indigenous Studies. Trent University, Peterborough, Ontario, May 2007.

Sherman, Paula. "The Friendship Wampum: Maintaining Traditional Practices in our Contemporary Interactions in the Valley of the Kiji Sìbì," in *Lighting the Eighth Fire: The Liberation, Resurgence, and Protection of Indigenous Nations.* Winnipeg: Arbeiter Ring Publishing, 2008.

Shuey, Chris. "Statement of Chris Shuey Before the Subcommittee on National Parks, Forests, and Public Lands Natural Resources Committee U.S. House of Representatives," March 28, 2008.

Stevens, Lawrence. "Testimony to Representative Grijalva's Subcommittee and the Subcommittee on Energy and Mineral Resources regarding Uranium Mining Issues in Northern Arizona," April 7, 2008.

Tauli-Corpuz, Victoria. "The Globalisation of Mining and its Impact and Challenges for Women," Paper Delivered at International Conference on Women and Mining, Baguio City, Philippines (January 1997).

Thompson, Ian, and Susan A. Joyce. "Mineral Exploration and the Challenge of Community Relations," *Report to The Prospectors and Developers Association of Canada* (1997).

U.S. Department of the Interior. "Cooperative Efforts Improve Watershed," Bureau of Land Management, U.S. Department of the Interior [homepage]; available from <www.blm.gov/wy/st/en/field_offices/Casper/lawncreek.html> accessed July 29, 2008.

U.S. National Park Service. "National Bighorn Canyon Recreational Area, Reclamation of Abandoned Uranium Exploration Sites," *Environmental Assessment (2003):* U.S. National Park Service [homepage]; available from <72.14.205.104/search?q=cache:jLtdEcyrhqwJ: www.nps.gov/bica/parkmgmt/upload/EAAML.pdf+environmental +assessment+policy+for+uranium+exploration+in+wyoming&hl= en&ct=clnk&cd=2&gl=ca> accessed August 2, 2008.

Wakelyn, Leslie. "Comments regarding EA0708-005—Bayswater Crab Lake," *Beverly and Qamanirjuaq Caribou Management Board* (May 23, 2008).

Whiteface, Charmaine. Personal Communication, March 23, 2008.

Wright, Philemon. An Account of the First Settlement of the Township of Hull on the Ottawa River, Lower Canada, delivered in 1823, (NAC FC 2495 H34 W75 1970z).

ENDNOTES

1 Omàmìwinini is a term in the language that describes our relationships as human beings within our homeland. It is also the term by which we are known in the spirit world. Many Omàmìwinini communities today use Algonquin as a replacement for the original term. I use both interchangeably in the book as a mechanism to reintroduce the term Omàmìwinini into our current consciousness.

2 Sherman, Paula. "The Friendship Wampum: Maintaining Traditional Practices in our Contemporary Interactions in the Valley of the Kiji Sìbì," in *Lighting the Eighth Fire: The Liberation, Resurgence, and Protection of Indigenous Nations.* Winnipeg: Arbeiter Ring Publishing, 2008.

3 Ibid.

4 The guiding principles collectively situate Omàmìwinini families within their traditional hunting areas that make up the community lands of Ardoch Algonquin people. The guidelines articulate Pimaadiziwin as the framework for interacting with our lands and waterscapes and provide principles that foster responsible behaviour in individuals and families. Within this framework, there are guidelines for development that promote human interests within the larger context of the Natural World. In this framework, development projects must be viewed within the larger ecological functioning of lands and waterscapes in our homeland. Any project that has the potential to alter the current state of ecological functioning should be reviewed carefully and possible

impacts identified. The position of the council since this time has been that the lands and waterscapes have already been impacted by four hundred years of development and must be maintained in their current state until enough time has passed to heal previous human activities.

5 Ardoch Algonquin First Nation, *Principles for Development.*

6 Ka-Pishkawandemin Family Head's Council Meeting. Ardoch, Ontario, November 26, 2006.

7 Reid, Christopher. "Legal Brief the Duty to Consult," prepared for the Uranium Defence website: <www.uraniumdefense.ca>.

8 Ibid.

9 Colour of right is a decision that belongs at a higher level of government than the police. It requires a decision about who has the right to access or occupy land.

10 See <www.attorneygeneral.jus.gov.on.ca/inquiries/ipperwash/index.html> for the transcripts to the inquiry.

11 Ibid.

12 Lovelace, Robert. "Testimony Before the Ontario Court," Kingston, Ontario, February 14, 2008.

13 Mary C. Hurley, *The Crown's Fiduciary Relationship with Aboriginal Peoples.* (Ottawa: Parliamentary Research Branch: Law and Government Division, 2000): pp. 1–11.

14 Philemon Wright, *An Account of the First Settlement of the Township of Hull on the Ottawa River, Lower Canada.* (NAC FC 2495 H34 W75 1970z).

15 Ibid.

16 Ibid.

17 Ibid.

18 Ibid.

19 "By the Chase" is a term that was often used by Europeans to describe the lifestyle of Indigenous peoples which they considered to be primitive and uncivilized.

20 Colonel Jarvis, Report. NAC RG 10, Vol. 186, pt 2, pg 108566I-108566K, Reel C-11, 511. See also, J.M. Higginson, Memorandum. NAC RG 10, Vol. 186, pt 2, pg 108, 566, Reel C-11, 511.

21 Ibid.

22 Ibid.

23 Some of those include the City of Ottawa, North Frontenac Township, Central Frontenac Township, South Frontenac Township, Beckwith Township, Lanark Highlands Township, Tay Valley Township, Lanark County, City of Kingston, Town of Perth, City of Peterborough, Highlands East Township, Town of Carleton Place, Township of Drummond/North Elmsley, Mississippi Mills, Haliburton County, Municipality of Dysart, Algonquin Highlands, Frontenac County, Minden, and the City of Kawartha Lakes.

24 C.A. Bate, Personal Communication, March 18, 2005.

25 Doug Brugge and Rob Goble, "The History of Uranium Mining and the Navaho People," *American Journal of Public Health* 92 no. 9 (2002): 1411.

26 Bruce E. Johansen, "The High Cost of Uranium in Navaho Land," *Akwesasne Notes* New Series 2 (April 1997): 10–12. See also Donald A. Grinde Jr. and Bruce E. Johansen, eds. *Ecocide of Native America: Environmental Destruction of Indian Lands and Peoples* (Santa Fe: Clear Light Publishers, 1994).

27 Department of Primary Industries, "Minerals & Petroleum," Department and Primary Industries Online [Homepage] available from <www.dse.vic.gov.au/DPI/nrenmp.nsf/9e58661e880ba9e44a256c640023eb2e/61839ca05e2f0250ca2573b600806acb/$FILE/ATTPZDJN/2%20GFEM%20-%20Drillholes.pdf> accessed March 3, 2008.

28 Charmaine Whiteface. Personal Communication, March 23, 2008.

29 Ibid.

30 Chris Shuey, "Statement of Chris Shuey Before the Subcommittee on National Parks, Forests, and Public Lands Natural Resources Committee U.S. House of Representatives," March 28, 2008. See also Lawrence Stevens. "Testimony to Representative Grijalva's Subcommittee and the Subcommittee on Energy and Mineral Resources regarding Uranium Mining Issues in Northern Arizona," April 7, 2008.

31 Giselle Lavalley, "Aboriginal Traditional Knowledge and Source Water Protection: First Nations' Views on Taking Care of Water," *A Report Prepared for the Chiefs of Ontario and Environment Canada* (March 2007): 9.

32 Ibid.

33 Ibid.

34 H.M. Salem. *Uranium Ores and the Environmental Impact on Human Health Risks* (Cairo: Cairo University Press, 2000): pp. 580–585.

35 Gordon Edwards. "Press Release in Support of Uranium Protest at Sharbot Lake," *Mining Watch Canada List Serve* Online [Homepage] available from <lists.miningwatch.ca/pipermail/news/2007-September/001555.html> accessed May 13, 2008.

36 Jonathan Samet and George R. Eradze. "Radon and Lung Cancer Risk: Taking Stock at the Millennium," *Environmental Health Perspectives* 108 Supplement 4: Occupational and Environmental Lung Diseases (August, 2000), pp. 635–641.

37 Ibid.

38 Samet and Eradze, 2000.

39 Ibid., 638.

40 Arizona Game and Fish Department. "Uranium Mining and Activities, Past and Present," *Update for the Arizona Game and Fish Department Commission.* Arizona: Arizona Game and Fish Department, May 2007. p 7.

41 Ibid.

42 Environment Canada, "Human Activities, Minerals Metals and Mining, Environmental Impacts of Mining Exploration," *Information Data Base: The State of Canada's Environment 1996* Environment Canada Online [homepage] available from <www.ec.gc.ca/soer-ree/English/SOER/1996report/Doc/1-7-4-7-6-2-1.cfm> accessed June 20, 2008.

43 Ibid.

44 Gerard Duhaime, Nick Bernard, and Robert Comtois. "An Inventory of Abandoned Mining Exploration Sites in Nunavik, Canada," *The Canadian Geographer* 49 no. 3 (2005): 262.

45 Ibid. Items and products observed on potential sites by key informants included:

Chemical products: Concentrated acids, sprays (aerosol, solvents), drums (brine of calcium chloride), battery acid, calcium chloride, dynamite, heavy metals.

Gas products: Propane tanks, propane

Oil products: Oil cans, diesel fuel, gasoline, oil, kerosene, furnace oil, standard 45-gallon drums

Buildings: Shacks, domes (metal), drilling cabins, garages, laboratories, mobile workshops, sheds (radio station), tripods

Dwellings: Cabins, mobile camp cabins, tent frames.

Scrap: Antennas, fasteners (nut, bolt, string), domestic appliances, tracked vehicle workshops, airplanes, boats, cans (food, milk), crushers, bulldozers, cables, trucks, pickups, camp equipment, canoes, car body (parts), wagons, loaders (tractor), containers, kitchen appliances, debris, derricks, shelves (drilling post), furnaces, generators, cranes, helicopters, beds, heavy equipment, airplane motors, outboard motors, snowmobiles, tools, water pumps, drums recycled as bridge, tracked vehicles (Bombardier, Muskeg).

Solid waste: Tent frames (collapsed), wood, boxes for rock samples, bottles, hoses, canvas, camp structures and equipment, rubber, plywood, domestic waste, dump sites, studs, mattresses, building equipment, food, seaplane wharves, Styrofoam, tents, cloth (plastic-coated), and sleighs (akutik).

46 Leslie Wakelyn, "Comments regarding EA0708-005—Bayswater Crab Lake," *Beverly and Qamanirjuaq Caribou Management Board* (May 23, 2008): 4.

47 Ibid.

48 Gary Scott, and Mark Wakeham. "Uranium Exploration in West Arnhem Land: A Report for the Environment Centre Northern Territory and the Australian Conservation Foundation," *Australian Conservation Foundation* (2001).

49 Ibid.

50 Ibid.

51 U.S. Department of the Interior. "Cooperative Efforts Improve Watershed," *Bureau of Land Management* U.S. Department of the Interior; available from <www.blm.gov/wy/st/en/field_offices/Casper/lawncreek.html> accessed July 29, 2008.

52 U.S. Department of the Interior, 2008.

53 U.S. National Park Service, "National Bighorn Canyon Recreational Area, Reclamation of Abandoned Uranium Exploration Sites," *Environmental Assessment (2003):* U.S. National Park Service [home page] available from <72.14.205.104/search?q=cache:jLtdEcyrhqwJ:www.nps.gov/bica/parkmgmt/upload/EAAML.pdf+environmental+assessment+policy+for+uranium+exploration+in+wyoming&hl=en&ct=clnk&cd=2&gl=ca> accessed August 2, 2008.

54 Ibid., 7.

55 Ibid., 12.

56 Ibid., 9.

57 Ibid., 12.

58 C. Paci, and N. Villebrun, "Mining Denendeh: A Dene Nation Perspective on Community Health Impacts of Mining," *Pimatisiwin: A Journal of Aboriginal and Indigenous Community Health* 3 no. 1 (2003): 79.

59 Ian Thompson, and Susan A. Joyce, "Mineral Exploration and the Challenge of Community Relations," *Report to The Prospectors and Developers Association of Canada* (1997): 4.

60 Ibid.

61 Thompson and Joyce, 5.

62 Ibid.

63 Ibid.

64 Ibid.

65 Iris Catholique, "Letter to Alan Ehrlich—Senior Environmental Assessment Officer," *Environmental Review.* Mackenzie Valley Environmental Impact Review Board (May 30, 2008).

66 Catholique, 2.

67 Susan A. Joyce, and Magnus MacFarlane. "Social Impact Assessment in the Mining Industry: Current Situation and Future Directions," *Mining, Minerals, and Sustainable Development* 46 (2001): 5.

68 Rabel J. Burdge, and Frank Vanclay, "Social Impact Assessment: A Contribution to the State of the Art Series," *Impact Assessment* 14 no. 1 (1996): 59.

69 Ibid., 60.

70 Ibid.

71 Gualnam Chinkhanmuan, "Mining Social and Environmental Impacts," *Asia Indigenous Peoples Pact Foundation* (April 30, 2008).

72 Ibid., 7.

73 Ricardo Carrere, *Mining: Social and Environmental Impacts.* (Montevideo, Uruguay: World Rainforest Movement, 2004), 7.

74 Ibid., 9.

75 Victoria Tauli-Corpuz, "The Globalisation of Mining and its Iimpact and Challenges for Women," Paper Delivered at International Conference on Women and Mining, Baguio City, Philippines (January 1997).

76 Ibid.

77 Ibid.

78 Burdge and Vanclay, 59.

79 Ibid., 61.

80 Ciaran O'Faircheallaigh. "Negotiating Cultural Heritage? Aboriginal–Mining Company Agreements in Australia," *Development and Change* 39 no. 1 (2008): 27.

81 O'Faircheallaigh, 27.

82 Ibid.

83 Ibid.

84 Ibid.

85 Ibid.

86 Adeline Jonasson, Chief. *North of 60 Mining News*, July 29, 2007; available from <www.petroleumnews.com/pntruncate/484739216.shtml> accessed August 1, 2008.

87 All Indian Pueblo Council, *Resolution 2007-12: Companion Resolution for the Protection of Mt. Taylor and all Sacred Sites and Cultural Properties Related to the Pueblos of Acoma and Laguna.* (New Mexico: All Indian Pueblo Council, June 2007): 2.

88 Ibid.

89 Jonasson, 2007.

90 Gregory Cajete, *A People's Ecology: Explorations in Sustainable Living* (Santa Fe: Clear Light Publishers, 1999): 3.

91 Ibid.

92 Ibid., 4.

93 Ibid., 6.

94 Cajete, 6.

95 Cultural documents are the material culture created by Indigenous peoples within their homeland. It includes such things as pictographs, petroglyphs, wampum belts, quill and beadwork, lodging, tools, and any other objects created as reflections of culture.

96 Cajete, 6.

97 Ibid.

98 Indigenous peoples as the centre of the wampum belt refer to teachings given by Elder William Commanda. The teachings are associated with the Friendship belt, which contains white and purple shell beads that are strung together to represent the relationships that Indigenous, French, and English people were to have with each other in what became Canada. An Algonquin figure sits in the centre of the belt with an English figure on one side and French on the other. The central figure represents the position that Algonquin people maintain as the people who have the responsibility of holding relationships together in our homeland.

99 Paula Sherman, "Indawendiwin: Spiritual Ecology as the Foundation of Omàmìwinini Relations," A Dissertation Submitted to the Committee on Graduate Studies in Partial Fulfillment of the Requirements for the Degree of Doctorate of Philosophy in the Faculty of Indigenous Studies. Trent University, Peterborough, Ontario, May 2007: 171.

100 Ibid.

101 Sherman, 2007, 171.

102 Ibid.